Barrier Islands

by ANNA PROKOS

With the Editors of TIME FOR KIDS

Table of Contents

Chapter 1
Islands That Protect Land

Have you heard of Hilton Head Island, South Carolina? What about Miami Beach, Florida, or Atlantic City, New Jersey? All these places are vacation spots. They are also **barrier islands**.

Barrier islands are long, narrow islands. They are between the ocean and the **mainland**. They keep strong winds and waves from sweeping onto the mainland. In this way, they are natural barriers, or blocks.

There are 405 barrier islands in the United States. A long chain of them runs down the East Coast, from Maine to Florida. The chain continues below Mississippi and Louisiana.

Myrtle Beach, South Carolina, with the Atlantic Ocean in the background

ME

NJ — Atlantic City

Outer Banks

NC

South Carolina — Topsail Island

Myrtle Beach

Charleston

Hilton Head

MS | AL | GA

LA

ATLANTIC OCEAN

FL

GULF OF MEXICO

Miami Beach

In this book you will see how weather and other forces form and shape barrier islands. You will also discover what can happen to a barrier island when people build houses, hotels, and roads there.

View from Above

Mary Edna Fraser is working to save barrier islands. Fraser is an artist in Charleston, South Carolina. She takes pictures of barrier islands from her own small plane. Then she uses the pictures to make artwork called batik. Batik is made by dying cloth. Fraser's artwork shows what barrier islands look like from above. She hopes that her artwork and a book she has written with Dr. Orrin H. Pilkey will make people want to save barrier islands.

Mary Edna Fraser takes a photograph from her plane.

This is an example of Fraser's batik artwork.

3

Chapter 2
Barriers and Building

Barrier islands are built by ocean activity. Moving ocean water, called a **current**, pushes sand toward the coast. A sandy shelf starts to form on the ocean floor. This is the start of an island. As more sand piles up, the island grows higher and higher.

| Ocean | Beach | Sand Dune | Trees and Shrubs | Salt Marsh | Bay | Mainla |

Different parts of a barrier island

Barrier Island Basics

There are certain things that almost all barrier islands have. They have a beach on the ocean side. They have a salt marsh on the mainland side. In between are sand dunes, trees and shrubs, and other natural environments. The marshes on barrier islands are home to sea animals and birds. Barrier islands are beautiful.

Sandy Shift

Another way barrier islands are alike is that they are always changing. They can change their size, shape, and location.

Over time, wind, waves, and **tides** wear down barrier islands. This is called **erosion**. Wind and water chip away at the ocean side of the island. The sand gets pushed to the middle and mainland side. The sandy shift can make barrier islands move as much as 6 meters (20 feet) each year. Old maps show that some barrier islands are miles away from where they once were!

Sometimes an island can change in only a few hours! Fast changes can be caused by **hurricanes**.

Shaping Up

Hurricanes, waves, and tides give barrier islands their shapes. Some islands, such as Little Corn Island, are shaped like chicken drumsticks. One end of these islands is swollen. The other end is narrow. Other long barrier islands are shaped like hot dogs.

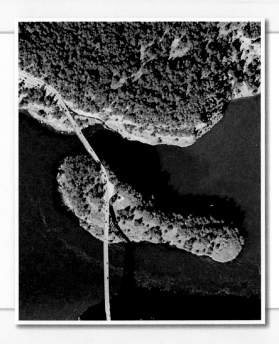

Seashore Cities

Many barrier islands have been turned into seashore cities. They have hotels, homes, businesses, roads, and everything else that comes with city living. This may be good for people. But it puts barrier islands at risk.

Buildings and roads are often built on the sands of barrier islands. To make room for them, the island's sand dunes are usually destroyed. The salt marsh may be filled in with sand or concrete. Animal and plant habitats, or homes, are often wiped out.

The buildings themselves begin to fall down, break, or wash into the ocean. No building can stay standing on ground that is moving all the time.

The Case of the Disappearing Island

There once was a barrier island called Dog Key. It was off the coast of Mississippi. In 1926, people changed the island's name to the Isle of Caprice. They built hotels there. Soon, the island began to shrink. Storms moved the sand. Plants that held the land in place had been taken away to make room for buildings. The sand washed away. By 1931, the island was completely underwater. Today, the island is just a memory.

Miami Beach, Florida, stands between the Florida mainland (left) and the Atlantic Ocean (right).

Chapter 3
Hurricane Hits

Barrier islands change. This fact can be seen during hurricane season. On the East Coast, hurricanes come in the summer and fall. When a hurricane hits a barrier island, three things happen. First, strong winds cause the **sea level** to rise. Water often floods a barrier island. This is called a **storm surge**.

Next, winds cause high waves to form on the surface of the ocean. Those waves move toward the shore and crash onto land. They can wash over all or part of a barrier island. The waves push large amounts of sand toward the back of the island. This can wipe out the island's salt marshes.

Folly Beach, South Carolina   Before Hurricane Hugo

Finally, heavy rains fall. This can cause floods. Large hurricanes can also flatten a barrier island. Anything that poked up from the sand before the storm—such as a house—may be flattened along with the island.

Barrier islands that have not been developed may bounce back from weather damage. This can take from a few weeks to a few years. For barrier islands that have been developed, it is another story.

Ocean waves can erode the shoreline on a barrier island.

Folly Beach, South Carolina   **After Hurricane Hugo**

Chapter 4
Stopping Nature

Buildings and roads can change a barrier island. During a storm or hurricane, large waves crash down on the beach. The waves wash away the front, or ocean side, of the island. Normally, waves and wind also place sand to the back, or mainland side, of the island. The front side is broken down. But the back side is built.

If buildings cover the island, they block the sand from being carried to the back side. The front of the island will continue to erode without the back part being built up. Over time, this may cause the island to disappear.

Some towns build sea walls as protection against powerful waves.

Sea wall

Building Walls

People try to stop barrier islands from changing shape. They build walls on the island to keep waves back.

Some towns dump large piles of sand on the island to slow down erosion. This works for only a short time. Then nature takes over.

Some towns put sand back onto a beach after a hurricane destroys it. They try to make the beach the same as it was before the storm. Yet this only solves the problem until the next big storm hits.

Piling up sand on a Florida beach

Topsy-Turvy on Topsail Island

In September 1996, Hurricane Fran swept across the ocean. It slammed into North Carolina. One of the hardest-hit spots was Topsail Island. This is a barrier island. Winds of up to 185 kilometers (115 miles) an hour pushed a storm surge across the island.

Large waves washed over parts of the island. Topsail's only highway was damaged. Some homes on the beach were destroyed. Other homes were swept into the ocean.

After the hurricane, the government changed the rules about building on Topsail Island. They said there were some areas where people could not build.

Barrier islands stand between storms and the mainland. They will always take a beating in a storm. An island in its natural state will be changed by a storm. But it can usually recover naturally. If an island is covered with homes and hotels, recovery will not happen naturally. It can be costly, too.

A Barrier Island in a Hurricane

- Most parts of barrier islands rise less than 5 meters (16 feet) above sea level.
- When hurricanes hit, waves surge over the height of the island.
- Plant and animal habitats on the island can be wiped out by storms.
- As a storm erodes the front part of an island, another part, usually the back, is often built up.
- It takes about two years for plants to grow on a barrier island that was washed over by a hurricane.

Homes along Topsail Beach, North Carolina, were destroyed by Hurricane Fran.

What Does the Future Hold?

Mary Edna Fraser and others are doing everything they can to save barrier islands. These islands help the mainland stand up to hurricanes and other storms. Barrier islands are home to many plants and animals. They are also a beautiful part of our country. With our help, they will stay a national treasure.

Charleston, South Carolina, with barrier islands in the distance

Glossary

barrier island (BAR-ee-uhr EYE-luhnd) narrow island that separates the ocean from the mainland *(page 2)*

current (KUR-uhnt) the movement of water in a certain direction *(page 4)*

erosion (i-ROH-zhuhn) when natural forces wear away the shore, roads, or buildings *(page 5)*

hurricane (HUR-i-kayn) a powerful storm with strong winds that forms over the ocean *(page 5)*

mainland (MAYN-land) the body of land forming the main or largest land mass of a region, country, or continent *(page 2)*

sea level (SEE LEV-uhl) the level of the world's oceans. It may rise or fall over long periods of time *(page 8)*

storm surge (STORM SURJ) water that swells under a hurricane at sea, often causing flooding of the shore *(page 8)*

tide (TIDE) the natural rising and falling of the surface of the ocean *(page 5)*

Index